THE HIGHWAYMAN

Alfred Noyes
illustrated by **Charles Keeping**

Oxford University Press

OXFORD   NEW YORK   TORONTO   MELBOURNE

The wind was a torrent of darkness among the gusty trees,
The moon was a ghostly galleon tossed upon cloudy seas.
The road was a ribbon of moonlight over the purple moor,

And the highwayman came riding—
        Riding—riding—
The highwayman came riding, up to the old inn-door.

He'd a French cocked-hat on his forehead, a bunch of lace at his chin,
A coat of the claret velvet, and breeches of brown doe-skin.
They fitted with never a wrinkle. His boots were up to the thigh.
And he rode with a jewelled twinkle,
        His pistol butts a-twinkle,
His rapier hilt a-twinkle, under the jewelled sky.

Over the cobbles he clattered and clashed in the dark inn-yard.
He tapped with his whip on the shutters, but all was locked and barred.
He whistled a tune to the window, and who should be waiting there

But the landlord's black-eyed daughter,
  Bess, the landlord's daughter,
Plaiting a dark red love-knot into her long black hair.

And dark in the dark old inn-yard a stable-wicket creaked
Where Tim the ostler listened. His face was white and
    peaked.
His eyes were hollows of madness, his hair like mouldy hay,
But he loved the landlord's daughter,
           The landlord's red-lipped daughter.
Dumb as a dog he listened, and he heard the robber say—

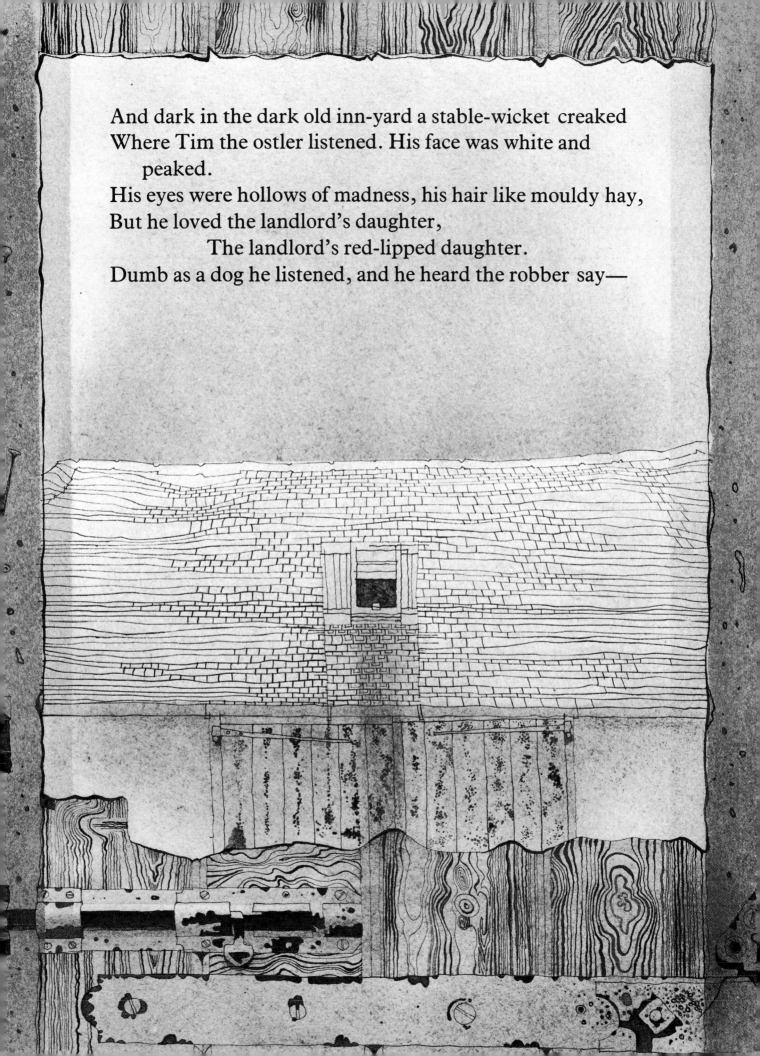

'One kiss, my bonny sweetheart, I'm after a prize to-night,
But I shall be back with the yellow gold before the morning light;
Yet, if they press me sharply, and harry me through the day,
Then look for me by moonlight,
             Watch for me by moonlight,
I'll come to thee by moonlight, though hell should bar the way.'

He rose upright in the stirrups. He scarce could reach her hand,
But she loosened her hair i' the casement. His face burnt like a brand
As the black cascade of perfume came tumbling over his breast;
And he kissed its waves in the moonlight,
      (Oh, sweet black waves in the moonlight!)
Then he tugged at his rein in the moonlight, and galloped away
   to the west.

He did not come in the dawning.

He did not come at noon;

And out o' the tawny sunset, before the rise o' the moon,
When the road was a gipsy's ribbon, looping the purple moor,
A red-coat troop came marching—
    Marching—marching—
King George's men came marching, up to the old inn-door.

They said no word to the landlord. They drank his ale instead.

But they gagged his daughter, and bound her, to the foot of her narrow bed.
Two of them knelt at her casement, with muskets at their side!

There was death at every window;
And hell at one dark window;
For Bess could see, through her casement,
the road that he would ride.

They had tied her up to attention, with many a sniggering jest.
They had bound a musket beside her, with the muzzle beneath her
 breast!
'Now, keep good watch!' and they kissed her.
  She heard the dead man say—
Look for me by moonlight;
  Watch for me by moonlight;
I'll come to thee by moonlight, though hell should bar the way!

She twisted her hands behind her; but all the knots held good!
She writhed her hands till her fingers were wet with sweat or blood!
They stretched and strained in the darkness, and the hours crawled by
 like years,
Till, now, on the stroke of midnight,
  Cold, on the stroke of midnight,
The tip of one finger touched it! The trigger at least was hers!

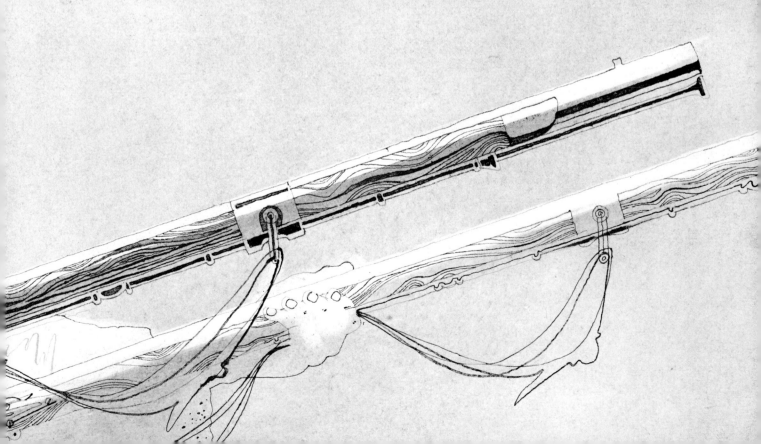

The tip of one finger touched it. She strove no more for the rest.
Up, she stood up to attention, with the muzzle beneath her breast.
She would not risk their hearing; she would not strive again;
For the road lay bare in the moonlight;
       Blank and bare in the moonlight;
And the blood of her veins, in the moonlight, throbbed to her love's refrain.

Tlot-tlot; tlot-tlot! Had they heard it? The horse-hoofs ringing clear;
Tlot-tlot; tlot-tlot, in the distance! Were they deaf that they did not
    hear?
Down the ribbon of moonlight, over the brow of the hill,
The highwayman came riding, Riding, riding!

The red-coats looked to their priming! She stood up, straight and still.

Tlot-tlot, in the frosty silence! Tlot-tlot, in the echoing night!
Nearer he came and nearer. Her face was like a light.
Her eyes grew wide for a moment; she drew one last deep breath,

Then her finger moved in the moonlight,
    Her musket shattered the moonlight,
Shattered her breast in the moonlight and warned him—
    with her death.

He turned. He spurred to the west; he did not know who stood
Bowed, with her head o'er the musket, drenched with her
own red blood!

Not till the dawn he heard it, and his face grew grey to hear
How Bess, the landlord's daughter,
        The landlord's black-eyed daughter,
Had watched for her love in the moonlight, and died in
the darkness there.

Back, he spurred like a madman, shouting a curse to the sky,
With the white road smoking behind him and his rapier brandished high
Blood-red were his spurs i' the golden noon; wine-red was his velvet
   coat;

When they shot him down on the highway,
        Down like a dog on the highway,
And he lay in his blood on the highway, with the bunch of lace at his
        throat.

And still of a winter's night, they say, when the wind is in the trees,
When the moon is a ghostly galleon tossed upon cloudy seas,
When the road is a ribbon of moonlight over the purple moor,

A highwayman comes riding—
Riding—riding—
A highwayman comes riding, up to the old inn-door.

Over the cobbles he clatters and clangs in the
    dark inn-yard.
And he taps with his whip on the shutters, but all
    is locked and barred.
He whistles a tune to the window, and who should
    be waiting there

But the landlord's black-eyed daughter,
      Bess, the landlord's daughter,
Plaiting a dark red love-knot into her
      long black hair.

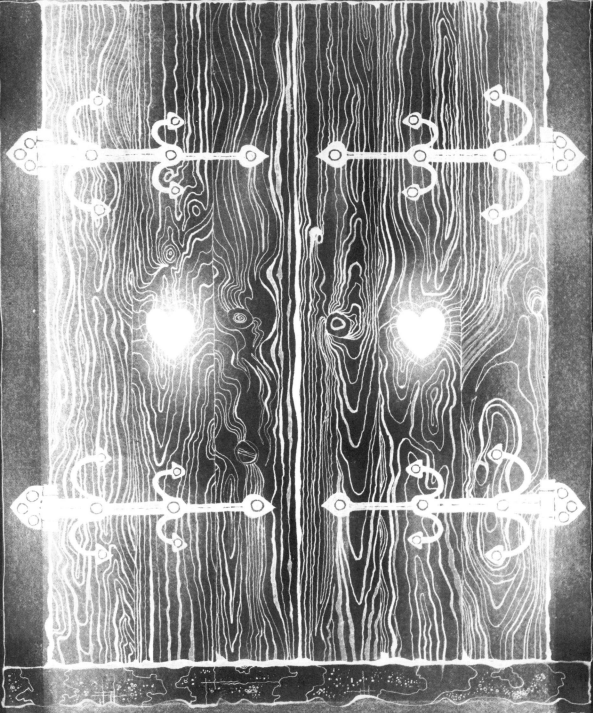

Oxford University Press, Walton Street, Oxford OX2 6DP

Text © Alfred Noyes 1913, renewed 1941
Illustrations © Charles Keeping 1981

First published 1981
Reprinted 1983 (three times), 1984, 1988, 1989, 1990, 1992, 1993
Softcover edition first published 1983
Reprinted 1983, 1984, 1988, 1989, 1990, 1992, 1993, 1994

Cataloguing in Publication Data

Noyes, Alfred
The highwayman.
I. Title II. Keeping, Charles
821'.912 PR6027.08H5 80–41855
ISBN 0–19–279748–4 (Hardcover)
ISBN 0–19–272133 X (Softcover)

Typeset in Great Britain
Printed in Hong Kong